Children's Environmental Health

2008
Highlights

*Environment, Health,
And a Focus on Children*

WHY FOCUS ON CHILDREN?

Since its founding in 1970, EPA's mission has been to protect human health and the environment. This report, eighth in an annual series from the Office of Children's Health Protection and Environmental Education, highlights the Agency's recent work on protecting the health of children by addressing the environments where they live, learn and play.

In the field of environmental protection, even experts did not always recognize that children are different from adults. Protective measures were first written with the average American adult—not children—in mind. Calculating the environmental contribution to disease is an evolving field, and the question of how much disease can be prevented through healthier environments is often asked. The World Health Organization estimates that one-quarter of the global disease burden is due to environmental factors. For children, that proportion rises to one-third. This burden is much greater in developing countries, where infant death from environmental causes is 12 times higher than in developed countries. Children encounter their environments differently than adults. Physically, their neurological, immunological, respiratory, digestive, and other physical systems are still developing and can be more easily harmed by exposure to environmental factors. Children eat more, drink more, and breathe more than adults in proportion to their body weight. Unclean food, water, and air therefore is more threatening to their health. Children also have unique exposure pathways, such as through the placenta or breast milk.

Children play and learn by crawling, placing their hands and other objects in their mouths—exploring the world with disregard for their own safety. These developmentally normal behaviors may lead to unintended exposure to environmental hazards. For example, a small child may be exposed to lead by putting fingers tainted with lead paint dust into her mouth. In addition, a child has little or no control over exposure, for example, to the second-hand smoke of a nearby adult. An older child might play on fields treated with pesticides, or sit day after day in a classroom contaminated with mold allergens.

PLACES AND SPACES WITH KIDS IN MIND

Brownfields: EPA's Brownfields Program began in response to community concerns about blight, disinvestment, and environmental contamination. In the 10-plus years since the program's inception, brownfields have been redefined. Properties that were shunned or eyed warily by those with the resources to transform them are now viewed as opportunities for environmental and economic rebirth. Communities with idle or abandoned properties have options and resources to turn them into cornerstones of positive change. EPA supports brownfields reuse rather than strictly enforcing for the environmental mistakes of a property's former owners.

Local health departments are eligible for EPA brownfields grants and may use up to 10 percent of their value to monitor the health of populations exposed to hazardous substances. Grantees can advocate for consideration of broader health and environmental issues related to site cleanup and urge sustainable redevelopment of brownfields. Grantees can expand parks, community gardens and exercise paths, thereby providing for children. www.epa.gov/brownfields

Our Town: Involving youth is a new component of the Brownfields Program. In 2003, EPA provided funds to Purdue University, in Lafayette, Indiana, to develop a program to educate youth about brownfields science and economics. The result was the Our Town program, managed by Purdue's Department of Engineering Education, which engages public school youth in brownfields-related activities, generating grassroots momentum for communities to actively pursue economic development and improve quality of life.

In Hammond, Indiana, more than 180 students from seven schools, in grades 4–12, helped identify 12 brownfields within the historically industrial city. Three of these properties were eventually selected for further investigation, assessments, and reuse planning. Students identified likely contaminants given the sites' histories, and confirmed the presence of underground storage tanks. They presented redevelopment options for the three properties at a community meeting, unveiling a plan to turn a former truck transportation property into a retail garden and landscape center. By the following school year, Our Town had been replicated in multiple schools in Kokomo, Lafayette, and Indianapolis. Our Town is now being implemented at schools in Chicago, Illinois; Philadelphia, Pennsylvania; Tucson, Arizona; and Portland, Oregon. Further expansion is expected.

From Brownfield to Children's Museum in Colorado: The city of Durango became home to the world's first steam-powered, alternating-current power plant in 1893. But the 8,000-square-foot site was boarded up in the 1970s and sat idle and dilapidated on the banks of the Animas River. In 2002, the Children's Museum of Durango wanted to restore and relocate to the site. Public health and environmental assessments were the first step, and the city and state worked with EPA's Brownfields Program on assessment and cleanup. Assessments found asbestos, pigeon waste, mercury, and uranium mining waste. Removal of asbestos and contaminated soil and a primary cleanup of the facility followed. Now, at the new Children's Museum, kids can experiment with the 19th century sciences that gave rise to electrical power, learn about locally mined energy, operate a hydrogen-powered race car, and explore energy-efficient building techniques.

Smart Growth and Children's Health: In 2004, EPA funded five community-based projects that make the connection between land use and children's health. Over the last few years, these grantees—ranging from WalkBoston to the Smart Growth for Healthy Children program in Cocke County, Tennessee—have begun to demonstrate positive impacts in their communities. These results represent a small but growing body of data and metrics that illustrate how, and to what extent, children can be active participants in improving the built environments where they live. See the *Journal of Public Health Management and Practice*, Volume 14 Number 3 (May–June 2008) for a summary of findings from these projects. www.epa.gov/smartgrowth

Vermont: Smart Growth Vermont developed the Healthy Kids/Healthy Neighborhoods program with assistance from EPA to engage youth in community planning and revitalization. By evaluating the health and safety of their neighborhoods and initiating service-learning projects that address their findings, youth play key roles in solving community issues and become excited about learning. Involving youth in decision making benefits everyone—youth, adults, and the community.

Massachusetts: The Concord River Greenway Park's goal is to restore, maintain, and enhance the ecological integrity and social viability of the Concord River as it flows through the city of Lowell. Through planning, the Concord River was transformed from a boundary between neighborhoods into a shared natural resource that unites and connects them to broader regional resources. Children were involved in the planning process and their input was invaluable in determining the scope and extent of the project.

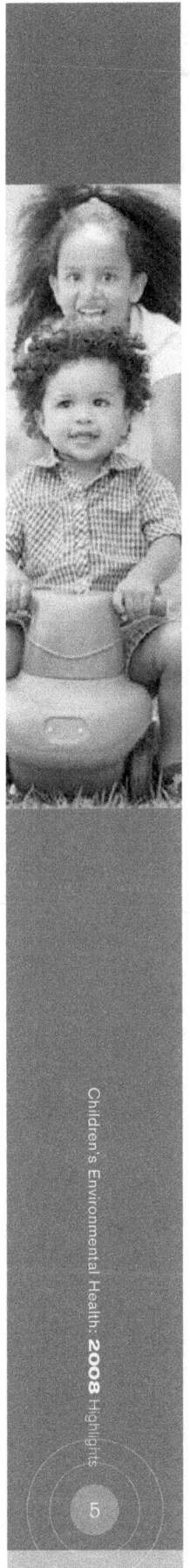

Smart Growth and School Siting:

EPA and the National Trust for Historic Preservation are looking at state policies that influence siting and construction of public schools. The aim is to identify and overcome state-level policy barriers to building community-centered schools, which allow children to walk to school and be physically active every school day. Community-centered schools also reduce emissions from automobiles and buses. As part of the project, the trust will work with five states to analyze policies, conduct outreach and education, create a plan for reform, and publicize the opportunities and benefits that accompany better school siting and construction.

Community Action for a Renewed Environment (CARE):

This grant program helps communities address multiple sources of toxic pollutants in their environments. EPA works with local organizations to set priorities for risk reduction and create self-sustaining, community-based partnerships that implement local solutions to problems posed by pollutants. Since 2005, EPA has provided nearly $8 million in grants to 48 communities in 25 states. Another 16 communities will receive grants in 2008. Many of these communities are focusing on children's health, making strides to reduce exposure to lead, mercury, and other chemicals. Among the many accomplishments: More than 1,300 homeowners received information and assistance on lead paint testing, 28 schools used EPA's Chemical Cleanout or Indoor Air Quality Tools for Schools (IAQ TfS) programs to eliminate chemical hazards, and the St. Louis, Missouri, city schools adopted no-idling zones to reduce exposure to vehicle exhaust. www.epa.gov/care

Examples of CARE Projects:

Denver: Through Ground-Work Denver and assistance from AmeriCorps, more than 100 at-risk 14 to17-year-olds designed and worked on a citywide environmental education program addressing children's health. They collaborated with students from the Community College of Denver to translate outreach materials on mercury hazards in the home into Spanish, and then distributed these materials. Students also conducted home visits in low-income communities to encourage parents not to smoke around children, and distributed EPA publications on children's environmental health.

Pennsylvania: Pennsylvania State University is partnering with local childcare centers and schools in Philadelphia to educate students, teachers, and maintenance staff about the benefits of Integrated Pest Management (IPM), reaching more than 1,000 people, garnering more than 225 pledges to use safer pest-control practices, and providing information in both English and Spanish.

Montana: Rocky Mountain College is working with partners to improve the health of American Indian children through chemical removal and disposal in schools and Boys and Girls Clubs, and by conducting outreach to K-12 schools on lead exposure, asthma triggers, and other environmental health issues. Visit http://cobalt.rocky.edu/~CARE/ to learn more about the successes of CARE in Indian Country.

CLIMATE CHANGE AND CHILDREN'S HEALTH

U.S. Climate Change Science Program (CCSP): The CCSP is a multi-agency effort focused on improving our understanding of the science of climate change and its potential impacts. Climate change, interacting with changes in land use and demographics, has the capacity to affect important human dimensions, especially human health, human settlements, and human welfare. A report by EPA, coordinated with CCSP, *Synthesis and Assessment Product 4.6: Analyses of the Effects of Global Change on Human Health and Welfare and Human Systems*, was released recently. www.climatescience.gov/Library/sap/sap4-6/final-report/

Climate change's projected effects include:

- greater frequency of heat waves
- increased variability of precipitation
- changes in minimum and maximum temperatures
- sea-level rise
- increased intensity of tropical storms
- more droughts
- more coastal and riverine flooding

Direct and indirect impacts to health and well-being can include morbidity and mortality associated with heat waves and extreme weather events, changes in the range and incidence of vector- and water-borne diseases, exacerbation of air pollution and aeroallergens, contaminated food and water, and impacts on mental health.

Children are more vulnerable to the health impacts of climate change, mainly due to physiological differences between children and adults. Children have a smaller body mass to surface area ratio than adults, making them more vulnerable to heat-related morbidity and mortality. They are more vulnerable to air pollution's health effects due to their increased breathing rates relative to body size. They spend time outside, often in active recreation, and their developing immune systems can make them more susceptible to water- and food-borne diseases. There is also some evidence that children are susceptible to psychological impacts, such as post-traumatic stress disorder, associated with extreme weather such as floods. See more information about children's health and climate change at http://yosemite.epa.gov/ochp/ochpweb.nsf/content/climate.htm

North Carolina Kids Fight Climate Change: EPA Region 4 (Alabama, Florida, Georgia, Kentucky, Mississippi, North Carolina, South Carolina, and Tennessee) presented a program to Durham Middle School on climate and energy. The school gathered more than 2,000 pledges to install 6,000 compact fluorescent lamps. The school's commitments are responsible for eliminating nearly 2.5 million pounds of carbon dioxide. The school has become an active environmental advocate and has formed an environment club. www.energystar.gov

New Jersey Earth Day: EPA Region 3 (Delaware, District of Columbia, Maryland, Pennsylvania, Virginia, and West Virginia) worked with third to sixth graders to explore how daily activities affect the environment. In the fifth-grade class of Hillside Intermediate in Bridgewater, New Jersey, 22 students decided to reduce the carbon footprint of their school by recycling and eliminating use of single-use plastic water bottles. The students developed a design for a reusable school water bottle. Proceeds from the bottle sales were used to buy recycling bins for the playgrounds. The class reduced the amount of single-use plastic water bottles students were using by 80 percent, resulting in savings of 26 gallons of gasoline, and one barrel of oil. The students plan to continue their efforts in the upcoming school year, with the goal of 100 percent use of the school water bottle. www.epa.gov/reg3esd1/childhealth/projects.htm.

PROTECTING CHILDREN WHERE THEY LEARN

Air Quality in Schools: According to a survey by the Centers for Disease Control and Prevention, more than half of the nation's schools are now implementing indoor air quality management programs. The study also determined that 86 percent of those schools based their efforts on EPA's IAQ TfS program, which is designed to reduce asthma attacks, bronchitis, and other respiratory ailments, as well as increase performance and productivity of students and staff. Key components of IAQ TfS include outreach and education, training, and recognition opportunities. www.epa.gov/iaq/schools

EPA has worked with the Connecticut Department of Public Health for the past five years to address indoor air quality by implementing and sustaining IAQ TfS in public schools. Partners have provided training to more than 500 schools, with a special focus on Hartford, New Haven, and Stamford. The project supports training workshops for custodians and provides Web-based tools for students, families, and staff. Nearly 90 percent of all public schools in Connecticut are participating, and reduced absenteeism and improved health in schools are well documented. In Vermont and New Hampshire, exemplary school districts are chosen as mentors to other school districts to implement TfS.

HealthySEAT: In 2007, EPA offered an enhanced version of its free, fully customizable Healthy School Environments Assessment Tool (HealthySEAT) to help school districts establish voluntary self-assessment and prevention programs covering every facet of school health and safety. www.epa.gov/schools/healthyseat

The Ohio School Environmental Health and Safety Inspection Rules, also known as "Jarod's Law," went into effect in September

2007. The law, named for a six-year-old killed in a school accident, is designed to protect and improve the environmental health and safety of Ohio's primary and secondary school students and staff. Jarod's Law requires experts from local health departments to annually inspect the school buildings and associated grounds within their jurisdictions to identify health and safety concerns. Recently, HealthySEAT was customized to include all Jarod's Law requirements. www.odh.ohio.gov/odhPrograms/eh/schooleh/sehmain.aspx

Clean School Bus USA: Every school day, 450,000 school buses transport 25 million students to and from school, sporting events, and other functions. Each year, those school buses emit tons of diesel exhaust that causes soot and smog and triggers asthmatic episodes. Since its launch in 2003, EPA's Clean School Bus USA program has worked to reduce children's exposure to diesel exhaust by reducing the pollution created by school buses. As a result, more than 2 million students are riding cleaner buses. An estimated 40,000 buses are using emission reduction technologies and cleaner fuels, and school districts across the United States are implementing idle-reduction programs, reducing air pollution and saving money on fuel. www.epa.gov/cleanschoolbus

> **Learning About Diesel:** The Clean School Bus USA program partnered with Scholastic Inc. to publish a new book in the popular *Magic School Bus* children's series. In *The Magic School Bus Gets Cleaned Up*, the children and Ms. Frizzle explore the pollution emitted from their own school bus and learn how to reduce the emissions as they travel through a diesel engine. The children learn about idle reduction and ways the community can help reduce health risks from diesel exhaust. At the end of the book, the magic bus gets its own pollution control device, a particulate-matter filter.

No Idle Zones: Idle-reduction programs have started statewide in Florida, North Carolina, and South Carolina. EPA Region 4 has supported more than 135 bus retrofit programs impacting more than 11,300 school buses, and supports wider use of biodiesel, hybrid, or electric school buses.

Better Buses: Students in Baltimore City Public Schools in Maryland will soon breathe cleaner air. An EPA grant will help the school system reduce diesel pollution by retrofitting 19 diesel-powered school buses with pollution control technologies. A second grant will support development of an indoor air quality program in 190 school buildings. Personnel at various levels will be trained to identify, recognize, and develop solutions to reduce exposure of building occupants to indoor pollutants.

Massachusetts Facility Administrators Association (MFAA): Over the past few years, MFAA, through a cooperative agreement with EPA's Region 1 (Connecticut, Maine, Massachusetts, New Hampshire, Rhode Island, and Vermont), has been developing and delivering an environmental health and safety training program for school facility managers across the state. Massachusetts is one of the first states to offer an accredited training program for school facility managers. The program received recognition during EPA's Environmental Merit Award ceremony on Earth Day 2008.

Healthier Illinois Schools: A new state law requires elementary and secondary schools, both public and private, to purchase and use only environmentally sensitive cleaning supplies. Only products that meet specific standards, such as those of EPA's Design for the Environment Program, may be used. This makes Illinois the second state (after New York) to require purchase and use of environmentally sensitive cleaning supplies in schools. EPA Region 5 (Illinois, Indiana, Michigan,

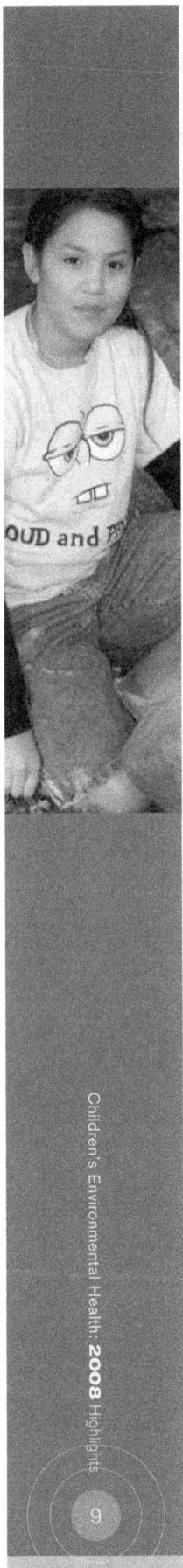

Minnesota, Ohio, and Wisconsin) helped plant some of the seeds of green cleaning in Illinois by funding a pilot project with the Healthy Schools Campaign, working with Chicago Public Schools, and supporting a sustainable schools project. www.standingupforillinois.org/green/school_cleaning.php

Chemical Cleanout Campaign: The Schools Chemical Cleanout Campaign inventories and removes outdated and potentially harmful chemicals from schools; conducts educational workshops for teachers on safe chemical management, lab safety regulations, and chemical waste disposal; and promotes green chemistry. www.epa.gov/sc3

In New Hampshire, Plymouth State University is conducting a Supplemental Environmental Project (an environmentally beneficial effort that a violator performs in lieu of payment of a portion of a penalty) to assist K-12 schools with chemical management practices, including staff training and waste removal.

Tribal Schools: EPA is reducing the health risks from exposure to hazardous waste, laboratory chemicals, pesticides, and asbestos by improving compliance at schools in Indian Country, where approximately 460,000 children and staff spend their days. These schools are often challenged by a need for ongoing and consistent environmental management and maintenance programs. EPA developed comprehensive, Indian Country–specific material and widely distributed the *Tribal Schools Compliance Assistance Notebook*. From 2005 to 2007, EPA conducted compliance monitoring inspections at 128 schools. A summary of EPA's efforts is available at www.epa.gov/compliance/data/planning/priorities/tribal.html, and the notebook is available at http://yosemite.epa.gov/R10/TRIBAL.NSF/Programs/Tribal+Schools.

Children's Health Initiative in Region 2: Through this four-year effort, EPA Region 2 (New Jersey, New York, Puerto Rico, and U.S. Virgin Islands) has protected 286,700 children from hazardous materials by helping schools identify and address environmental concerns. Through integration of the drinking water, asbestos, pesticides, and hazardous waste programs, EPA is focusing on K-12 school systems in specific communities to ascertain the status of testing drinking water for lead in schools and to review compliance with hazardous waste regulations. In Trenton, New Jersey, EPA collected 568 drinking water samples from 10 elementary schools. Test results, then remediation, are the next steps. www.epa.gov/region02/children/index.html

ASTHMA

Asthma is a complex chronic disease affecting millions of people in the United States, including 6.5 million children. The burden of asthma is great: Emergency room and hospital visits, missed school days, and missed work days, all contribute to an estimated annual national cost of $19 billion.

Communities Addressing Asthma: 290 community-based asthma programs have joined the EPA Communities in Action for Asthma Friendly Environments Network. Participants pursue strategies to achieve positive health outcomes, including cultivating program leaders, establishing sound community relationships, maximizing cooperative opportunities, providing integrated health care services, and implementing tailored environmental interventions. The network is supported by an interactive Web site that provides education, communication, and resource sharing. The network and supporting organizations

meet annually at the EPA-sponsored National Asthma Forum, which provides events throughout the year for community leaders to accelerate adoption of best practices for asthma care. This year, a pre-forum workshop focused on building a business case for comprehensive asthma management that includes patient education and environmental management. www.asthmacommunitynetwork.org

Heartland Tackles Asthma: EPA Region 7 (Iowa, Kansas, Missouri, and Nebraska), in partnership with the American Lung Association of the Central States, hosted

the first-ever Heartland Region Asthma Forum, in Omaha, Nebraska. A comprehensive agenda covered the new National Guidelines for the Diagnosis and Management of Asthma, EPA Communities in Action for Asthma Friendly Environments initiative, and workshops on indoor environmental assessments and outdoor air quality.

Southwest Asthma Advances:

Through grants, organizations throughout EPA Region 6 (Arkansas, Louisiana, New Mexico, Oklahoma, and Texas) are training children and their families to reduce environmental asthma triggers. In Oklahoma, 186 children were trained on how to reduce environmental asthma triggers through the Pawnee Health Clinic. Along the Texas/Mexico border, more than 75 health Promotoras were trained. In Arkansas, New Mexico, and Texas, 17 school nurses, 72 families, 44 health care professionals, and 211 child care providers have been trained this year. In rural East Texas, 30 children were treated by the Region 6 Pediatric Environmental Health Specialty Unit (PEHSU)—the Southwest Center for Pediatric Health—in the Breath of Life, a mobile asthma education and treatment van.

Arresting Asthma Triggers: Make the

Road by Walking is a community organization in Brooklyn, New York, that is raising awareness about children's asthma triggers. EPA provided a cooperative agreement to benefit the Bushwick neighborhood, where children suffer from the old housing stock and lack of information on how to avoid exposure to asthma triggers. Many of the residents are renters and require landlords to make repairs and maintain units. The project reduced exposure to asthma triggers by educating more than 1,200 residents, providing extended hours for asthma treatment services, expanding local partnerships, and convening a new coalition of 10 community groups to influence policy. www.epa.gov/compliance/ environmentaljustice/grants/ej-cps-grants.html

Innovative Asthma Programs: In

2008, EPA selected three organizations for the National Environmental Leadership Award in Asthma Management. The award recognizes health plans and health care providers who have demonstrated leadership in managing environmental triggers as part of a comprehensive asthma management program, and who are improving the health of children they serve. Award winners serve as mentors and models for best practices in asthma care. www.epa.gov/asthma

2008 Winners are:

Monroe Plan of Rochester,

New York: The Monroe Plan covers 5,633 children with asthma in Monroe County and 12 neighboring rural counties. In partnership with ViaHealth, a health care delivery system, the plan launched a program to shift asthma care away from emergency services and inpatient care and toward improved patient self-management. The program now covers all plan members with moderate to severe pediatric asthma and includes assistance to health care providers. Asthma action plans, education, and home assessments helped reduce emergency room visits from 1.1 per person to 0.95 visits per person over the first three years of the program. Inpatient admissions decreased from 98.3 per 1,000 to 84.15 per 1,000 in the first three years.

University of Michigan Health System, Ann Arbor, Michigan:

This health system is a nonprofit health care provider serving 12,214 adults and children with asthma in four counties in southeastern Michigan. The program includes in-home asthma education through the Michigan Visiting Nurses Asthma Home Environmental Assessment Program. This award winner achieved a 50 percent decrease in asthma-related hospitalizations from July 2005 to June 2007. Between June 2006 and June 2007, the program had a 60 percent decline in emergency room visits and 85 percent fewer hospitalizations.

The **Asthma Network of West Michigan**, in Grand Rapids, is a community coalition that provides comprehensive home-based case management to 94,500 children and adults with asthma in western Michigan. Results include improved health outcomes and cost savings. This success has led to a partnership with Priority Health (a winner of the 2007 National Environmental Leadership Award in Asthma Management), which agreed to reimburse the network for its home visit program. The partnership is the nation's first between a grassroots coalition and managed care plan. The network has contracts with five local health plans, and its asthma management program provides asthma education, coordination with health care providers, development of asthma action plans, home environmental assessments, and social service support. This comprehensive care has led to a 64 percent decrease in hospitalizations and 60 percent decrease in emergency room visits. These improved health outcomes resulted in approximately $800 in net health care cost savings per child per year.

Goldfish Media Campaign: The National Childhood Asthma Public Service Media Campaign, a collaboration between EPA and the Ad Council, uses the imagery of a goldfish out of water to convey the message of urgency and provides tips for parents to help manage asthma triggers, such as making their homes smoke-free and controlling mold and dust mites. The Ad Council recently conducted a study to determine the ads' effectiveness. The campaign outscored

commercial advertising on measures of consumer likability; engaging and empowering families, especially African-American and Hispanic families; promoting asthma management behaviors; calls to the hotline; and visits to the Web site for asthma education information. The Chris Draft Foundation joined the goldfish campaign in 2008, and Draft himself, a professional football player, is featured in radio announcements. www.noattacks.org

Asthma Health Outcomes Project: The project is a landmark study sponsored by EPA to understand asthma disparities and determine how communities can best improve health outcomes of people with asthma. EPA and the Centers for Disease Control and Prevention's National Center for Environmental Health (NCEH) coordinated a successful half-day workshop at Meharry Medical College in Tennessee and a seminar at Morehouse School of Medicine in Georgia, where more than 185 faculty, students, and health care professionals were educated about asthma.

SAFE IN THE SUN

SunWise Communities: With more than 1 million people diagnosed with skin cancer each year, teaching children to be safe in the sun is an important step toward decreasing the incidence of this preventable disease. Since sun damage is cumulative and protection is best if started early, EPA's SunWise Program works to improve children's sun safety knowledge, attitudes, and behavior. A recent study published in *Pediatrics* (vol. 121 no. 5) found that SunWise efforts between 1999 and 2015 should prevent more than 50 premature deaths and 11,000 cases of skin cancer in the United States, resulting in a $2–$4 savings in public health costs for every federal dollar invested. In addition to the more than 21,000 schools and informal education organizations now signed up with SunWise, three more areas became SunWise Communities this year. www.epa.gov/sunwise

- The city of Boston has pledged to plant 100,000 trees by 2020, increasing its canopy cover by 60 percent.

- Recent statistics on skin cancer indicate that Cobb County, Georgia, has a 51 percent higher incidence of melanoma than the national average. In response, the county has distributed SunWise kits to all its schools and provided education about overexposure to ultraviolet rays.

- Washington State built on previous SunWise efforts in King and Pierce Counties to become the third state to push for sun safety in its schools. Washington has the fifth highest melanoma incidence rate and seventh highest death rate in the country. Melanoma occurs almost as frequently in the cloudy and rainy part of the state as the sunny and dry eastern part.

THE AIR THEY BREATHE

Ozone Standard: On March 12, 2008, EPA announced a new eight-hour standard for ozone. The new primary eight-hour standard, which sets limits to protect public health, including the health of sensitive populations such as asthmatics, children, and older adults, was changed from 0.080 parts per million to 0.075 parts per million. Benefits for children's health include preventing some cases of bronchitis and aggravated asthma, decreasing hospital and emergency room visits, and reducing school absenteeism due to respiratory illness. www.epa.gov/groundlevelozone

Smoke-Free Homes for Head Start Children: Exposure to secondhand smoke can cause middle ear infections, bronchitis, pneumonia, decreased lung function, sudden infant death syndrome, and can worsen asthma. A 2007 Surgeon General's report, reaffirmed that there is no safe level of secondhand smoke. Millions of young children continue to be exposed to it in their homes.

EPA and the U.S. Department of Health and Human Services (DHHS) are improving the quality of life for nearly 1 million Head Start children through secondhand smoke and asthma outreach efforts. Head Start teachers, staff, and parents will create smoke-free, asthma-friendly homes using tools and resources from the agencies. www.epa.gov/smokefree

Residential Wood Smoke: EPA's Great American Woodstove Changeout campaign is facilitating replacement of older, dirtier woodstoves and fireplaces with new, cleaner burning appliances such as masonry heaters and gas-fueled, pellet-fueled, and EPA-certified stoves. Particle pollution, such as the particles in wood smoke, can be harmful to children. Since 2005, the program has replaced more than 7,500 woodstoves and fireplaces, avoiding nearly 200 tons of particle pollution and realizing an estimated $50 million per year in health benefits. www.epa.gov/woodstoves

Outdoor Wood Heaters: EPA has also developed a voluntary program that is helping to bring cleaner outdoor wood heaters to market. These heaters, also known as outdoor wood boilers, look like a small shed with a smokestack, typically located on the outside of the building to be heated. They burn wood to heat water that is piped underground to provide heat and hot water to homes, barns and greenhouses. The heaters are mostly used in rural, cold climates where wood is readily available; however, they can be found throughout the United States. EPA encourages manufacturers to produce cleaner models, encourages users to buy the cleanest models available, and educates users on the health effects of woodsmoke. New heaters that qualify for EPA's program are at least 70 to 90 percent cleaner than existing units. www.epa.gov/woodheaters

Reducing Children's Exposure to Household Smoke: More than half the world's population relies on solid fuels (e.g., wood, dung, crop residues, and coal) for everyday cooking and heating needs, filling homes with harmful particulate matter, carbon monoxide, and toxic smoke. Among children younger than five, breathing unsafe levels of this smoke more than doubles the risk of acute lower respiratory infections. Worldwide, such infections continue to be the biggest killer of young children, causing more than 2 million deaths annually and accounting for 19 percent of deaths in children under five. The Partnership for Clean Indoor Air, launched by EPA in 2002, is tackling this issue by introducing safer, cleaner burning, fuel-efficient cooking and heating technology. Today, more than 170 public and private sector partners are working in 67 countries to improve health by reducing household air pollution. This effort has helped more than 1.4 million households adopt clean and efficient cooking and heating practices. Partners aim to reach another 6 million households by 2010. www.pciaonline.org

PROTECTING CHILDREN FROM PESTS AND PESTICIDES

Public Health and Urban Pests: The spread of mosquito- and tick-borne diseases in Europe and North America and the asthma epidemic in industrialized nations have signaled the need to carefully assess the threat of urban pests to public health. According to the World Health Organization (WHO), asthma attacks are triggered by an allergic response in 50 percent of adults and in 80 percent of children with asthma. In urban areas, sensitization to pests, including rodents, cockroaches, and dust mites, is common among asthmatics. EPA supported a WHO project in which international experts in pest-related fields provided evidence for policies to address this public health problem. These experts identified the public health risks posed by various pests and measures to prevent and control them, and in 2008 WHO published the book *Public Health Significance of Urban Pests* (www.euro.who.int/document/e91435.pdf).

Integrated Pest Management (IPM) Promotes Healthier Schools: EPA recommends that schools use IPM to reduce pesticide exposure to children. IPM is a safer and usually less costly option than conventional pest management. IPM programs use common sense strategies to reduce sources of food, water, and shelter for pests in and around buildings, and encourage careful use of pesticides. More than 20 percent of schools use IPM. EPA's goal is to have all schools on board by 2015. www.epa.gov/pesticides/ipm

In Indiana, the Monroe County Community School Corporation's IPM program was developed in the mid-1990s and has been replicated in 10 states. This program boasts a 90 percent reduction in pesticide use, pest problems, and pest-control costs.

In Utah, the Salt Lake City School District became the nation's 28th school district to earn the IPM Institute of North

America's IPM STAR® certification following a rigorous, 37-point inspection. Both pesticide applications and pest occurrences have dropped by more than 90 percent since the district's program began. EPA Region 8 (Colorado, Montana, North Dakota, South Dakota, Utah, and Wyoming) provided funding for the effort.

Integrated Pest Management (IPM) In Section 8 Housing: A partnership between EPA and the Department of Housing and Urban Development is bringing IPM to families living in Section 8 housing (a federal program that provides housing assistance to eligible low-income renters and homeowners) through a pilot program called the Market-to-Market Green Incentives Initiative. This program provides financial incentives for property owners to green their rehabilitation and maintenance practices, including a requirement for switching from conventional, chemical-focused pest management to prevention-oriented IPM strategies. The program is expected to reach about 2,400 Section 8 developments over the next five years, involving about half a million low-income residents. The potential health benefits of reducing exposure to pests and pesticides are significant—50 percent of residents report at least occasional cockroaches or rodents and 10 percent report always having cockroaches or rodents. Research into allergic asthma shows a strong correlation between allergies to cockroaches and asthmatic response. Most allergic asthma sufferers are children in low-income urban areas.

Reducing Pesticide Exposure in Child Care Centers: This year, EPA and DHHS' Office of Head Start launched a national pesticide awareness campaign called Play It Safe. The campaign features a suite of outreach materials, available in English and Spanish, targeting Head Start parents and staff. Training for 250 child-care employees serving more than 1,500 children was provided in 2008 by the Pennsylvania Integrated Pest Management Program (PA IPM). PA IPM is training staff of Philadelphia's Latino child care facilities.

Recognizing IPM Leaders: EPA supports the IPM Institute's IPM STAR certification program, which provides incentives to reduce pests and pesticide risks to children's health. Certification clearly establishes competence in a way that is recognized by others. Applicants undergo careful on-site evaluations of their IPM programs. To date, practitioners in more than 35 school districts and child-care centers are certified and more are waiting to go through the process. http://ipminstitute.org/school.htm

Safer Rodent Control: In May, EPA announced new safety measures for 10 rodent-control products. Rodent-control pesticides, or rodenticides, are an important tool for public health pest control, including controlling mice and rats around the home. However, these products have been associated with accidental exposures to thousands of children each year. Rodenticides marketed to consumers must be enclosed in bait stations, making them inaccessible to children. Loose bait, such as pellets, is prohibited for use in homes. Data indicate that children in low-income families are disproportionately exposed to rodenticides. http://epa.gov/pesticides/reregistration/rodenticides/

Latino Outreach to Prevent Pesticide Poisoning: An outreach campaign during National Poison Prevention Week targeted Latino families and reached 32 million people in the United States and Latin America with the message "Children act fast, and poisons do, too!" American Association of Poison Control Centers (AAPCC) data show that more than 50 percent of the 2 million incidents of exposure to chemicals and other materials each year involve children younger than six, with 90 percent of calls concerning home exposures. EPA's Pesticides Hispanic Outreach Initiative reduces exposure risk by showing how to minimize exposure, defining the symptoms of pesticide poisoning, and providing information on where to get help.

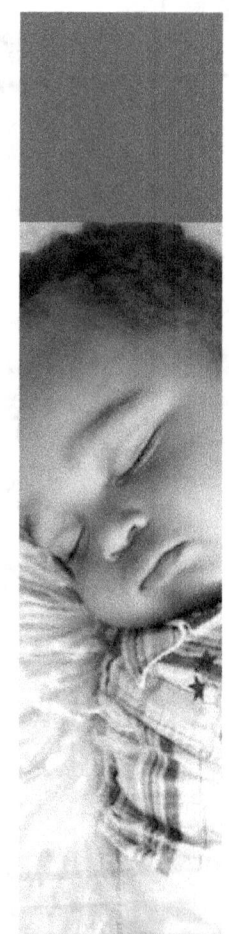

KEEPING CHILDREN LEAD-FREE

Reducing Lead Poisoning: Children are exposed to lead most often through unintentional ingestion of lead-containing particles, such as paint dust or contaminated soil. Under the 1992 Residential Lead-Based Paint Hazard Reduction Act, EPA established training and certification rules and standards for lead in paint, dust, and soil. In April 2008, EPA issued a new rule aimed at protecting children from lead paint hazards. The Lead Renovation, Repair, and Painting Program rule (40 CFR Part 745) requires contractors and construction professionals to be certified and to use lead-safe work practices during renovation, repair, and painting in pre-1978 housing and child-care centers, pre-schools, and kindergartens. The rule also requires contractors to provide a new lead hazard information brochure to property owners, tenants, and owners and operators of buildings that have child-occupied facilities, as well as to the parents and guardians of children under age six using the facilities.

The rule will be fully effective in April 2010.
www.epa.gov/lead/pubs/renovation.htm

Lead Grants:

- In 2007, EPA provided more than $3 million through the National Community-Based Lead Grant Program to educate those at risk, provide lead awareness training, and develop local ordinances aimed at lead abatement in communities with older housing. Grant recipients include city health departments, colleges and universities, community organizations, religious groups, and other nonprofit organizations.

- The Targeted Lead Grant Program funds projects in areas with high numbers of children with elevated blood lead levels. In 2007, EPA awarded $5.2 million in grants to address immediate needs of communities and highlight model lead-poisoning prevention strategies.

- EPA awarded nearly $1 million in grants to 15 tribes to reduce the incidence of childhood lead poisoning and support educational outreach and baseline assessments of exposure. www.epa.gov/oppt/lead/

Less Lead, More Water: Drinking water is also a source of lead exposure in children. EPA is addressing lead in drinking water in school and child-care facilities by developing tools and guidance. These materials encourage voluntary lead reduction programs and help school officials and child-care providers minimize lead in drinking water. To date, more than 1,300 toolkits, 730 school guides, 3,900 child-care facility guides, and 1,300 DVDs have been distributed. www.epa.gov/safewater/schools

Getting the Lead Out of Pacoima: The community of Pacoima, California, used a $100,000 grant from EPA to reduce exposure to lead paint and increase blood lead testing for children at risk. Many of the 22,035 homes in this high-density, low-income, mostly Latino community are more than 30 years old and still contain lead paint. Pacoima Beautiful, a community organization, partnered with local organizations and trained them to work with residents to raise awareness of the risks of lead paint, the importance of lead testing in children, and the need for home testing and abatement. The program tested 675 children for blood lead levels, provided information to 2,500 residents on safe cleaning practices and other simple measures to reduce lead levels, and tested 300 homes for contamination. Of those homes, 31.4 percent exceeded the lead dust criteria for floors and windows. Of those that exceeded the criteria, 27 percent have been renovated or referred to free services or low-interest loans to remove the lead hazards. www.epa.gov/compliance/environmentaljustice/grants/ej-cps-grants.html

Boston, Massachusetts, Lead Campaign: EPA Region 1 (Connecticut, Maine, Massachusetts, New Hampshire, Rhode Island, and Vermont) seeks to end childhood lead poisoning in Boston by 2010. The Lead Action Collaborative includes Tufts University and other public, private, nonprofit and housing organizations and increases visibility of the issue, creates strong lead policies and regulations, and targets enforcement and outreach to reduce risk to children. This coordinated effort has produced impressive results: The number of lead-poisoned children in Boston dropped from 1,123 in 2001 to 362 in 2007.

Flakes Play in North Carolina: The Durham Affordable Housing Coalition conducted a public education project, Lead Safe...For Our Children's Sake, to reduce childhood lead poisoning. Targeting African-American and Latino families living in high-risk homes, this effort resulted in "Flakes Play," a musical production about lead-poisoning prevention produced by the Walltown Children's Theater. The play has heightened public awareness of the dangers of lead and how to prevent lead poisoning in children, especially in vulnerable, low-income families. www.epa.gov/lead/pubs/lppwregion.htm#4

Toxicity and Exposure Assessment for Children's Health (TEACH):

TEACH is EPA Region 5's (Illinois, Indiana, Michigan, Minnesota, and Wisconsin) 10-year effort to consolidate and summarize current peer-reviewed scientific literature published since 1972. TEACH focuses on a subset of chemicals relevant to children's environmental exposure and toxicity. The project is Web-based and has two major components: a searchable database and chemical summaries. The database is used to search summaries of literature on children's exposure and toxicity. The chemical summaries highlight information from the database and other sources, such as EPA's Integrated Risk Information System. TEACH currently provides information for 18 chemicals or chemical groups, including arsenic, benzene, formaldehyde, three forms of mercury, and polychlorinated biphenyls (PCBs). www.epa.gov/teach

Children's Exposures to Chemicals In Their Everyday Environments:

Researchers in EPA's National Exposure Research Laboratory continue to work to develop, refine, and validate analytical methods for different classes of chemicals found in children's everyday environments, including brominated flame retardants, perfluorinated chemicals, and current-use pesticides. A number of manuscripts have been published in peer-reviewed literature, including articles on methods for measuring pesticides in house dust, lead in paint, and perfluorinated chemicals in house dust, soil, water, and fish. Data analysis has been performed and published in the scientific literature for a number of recent studies. Recently published manuscripts have reported on sources and pathways of exposure, important exposure factors, and relationships of environmental measurements to biomonitoring results. EPA researchers are collaborating with academics and other federal agencies to collect and analyze data on children's exposures. These efforts allow EPA scientists to understand what chemicals children are exposed to in their everyday environments, what the sources of those chemicals are, what the most important exposure factors are, and how exposures and risks can be reduced. www.epa.gov/nerl

Mercury, PCBs, and Fish in New York and New Jersey:

An estimated 16 percent of women of reproductive age in the United States eat fish at least once per day. While a good source of protein, studies show some fish and shellfish contain mercury, PCBs, pesticides, and other harmful contaminants at levels that can result in pre- or postnatal impairments. New York City's Health and Nutrition Examination Survey biomonitoring study revealed that New Yorkers have more than three times the national average of mercury in their blood. One-quarter of the 1,811 New York residents tested had blood mercury concentrations at or above the 5 µg/L New York State reportable level. The study also found that foreign-born Chinese residents had blood mercury concentrations more than two and a half times that of the general population in the city. Although the fish species commonly associated with the highest levels of mercury are not seen frequently in Chinese kitchens, the study found that Chinese-Americans eat fish three times more often than others in the city. www. ehponline.org/docs/2007/10056/abstract.html

In response to the findings, EPA funded two public health projects. The first project will develop a geographic information system (GIS) tool to identify areas in New York and New Jersey where women of child-bearing age are at highest risk of eating contaminated fish. The second project will determine mercury and PCB levels in fish species most commonly sold in the New York City area. This will provide an effective way to reach vulnerable populations. The second project involves fish testing from the largest wholesale market in the United States. An exciting and innovative aspect of the testing is using DNA coding for accurate species identification. Both studies are anticipated to improve targeting of future outreach efforts.

Reducing and Eliminating Mercury Use and Exposure:
Mercury is contained in some products, such as thermostats, thermometers, and barometers, that are commonly found in homes, health care facilities, and schools. If these products break, toxic mercury vapors can be inhaled. EPA is taking action to reduce and eliminate mercury in products by developing information on safer alternatives, working with states to reduce use of mercury in products and promote mercury-containing product collection and recycling, developing regulations to reduce re-introduction of discontinued products, and working with other federal agencies on short- and long-term management of surplus elemental mercury supplies. EPA is also working with other countries and the United Nations Environment Program to address risks associated with mercury uses, releases, and exposure internationally. EPA is working with Argentina, China, Costa Rica, and Mexico to reduce mercury in hospitals; and with Burkina Faso, Chile, Ecuador, Mexico, Panama, and South Africa to conduct mercury product inventories, market studies, and risk management projects. http://epa.gov/mercury/

Fish Consumption Information:
EPA's National Fish Advisory Program recently released a new Fish Kids Web site (www.epa.gov/fishadvisories/kids/) that teaches kids about contaminants in fish and fish consumption advisories using fun, interactive stories and games. The Web site is designed for kids ages 8–12 and is best used with an adult. Whether they buy fish or catch their own, kids and their families can use this site to learn how to choose fish wisely.

Understanding Chemical Exposures:
EPA's Voluntary Children's Chemical Evaluation Program (VCCEP) is designed to evaluate hazards, exposures, and risks of chemicals to children, and to develop information needed to assess these risks. Companies that manufactured or imported 20 of 23 chemicals to which children have a high likelihood of exposure have committed to provide EPA with information on health effects, exposure, risk, and data needs. The information provided to date on 15 chemicals, their subsequent reviews, and related Agency decisions are provided on the Web site (www.epa.gov/oppt/vccep/). After conducting an interim evaluation of the program, EPA is considering using a modified VCCEP approach to follow up on chemicals of interest in its new Chemical Assessment and Management Program, which assesses and initiates action on more than 6,750 industrial chemicals manufactured in quantities of 25,000 pounds or more. www.epa.gov/champ/

ENVIRONMENTAL HEALTH INDICATORS

America's Children and The Environment is a leading source of information on environmental factors related to the health and well-being of children in the United States. The Web site and reports present data on levels of contaminants in the environment that can affect children's health, and on childhood illnesses that can be influenced by exposure to contaminants. Data are updated annually, most recently in July 2008. www.epa.gov/envirohealth/children

Some examples:

• Approximately 53 percent of U.S. children lived in counties where ozone levels exceeded National Ambient Air Quality Standards at least one day in 2006. (see chart below)

• About 10 percent of children were served by community water systems that exceeded a Maximum Contaminant Level or violated a treatment standard in 2006.

• The median concentration of lead in the blood of children five years old and younger dropped from 15 micrograms per deciliter (µg/dL) in 1976–1980 to 1.6 µg/dL in 2003–2004, a decline of 89 percent.

• In 2006, 9.3 percent (6.8 million) of all children had asthma.

International Children's Environmental Health Indicators:

EPA has been working with WHO and other organizations to promote development of children's environmental health indicators internationally, an effort initiated at the World Summit on Sustainable Development in 2002. The pilot phase of this work culminated in an international workshop in Tunisia in April 2008. Participants reviewed and assessed the progress made on developing the health indicators based on reports prepared for North America, Europe, Argentina, Cameroon, Kenya, Oman, Tunisia, and Zimbabwe. Key conclusions from the workshop focused on next steps for improving and expanding children's environmental health indicators, and reflected a strong commitment by all participants to continue this work. The consensus of the workshop was that a small "core set" of indicators, representing common interests and concerns across regions and nations, should be further developed by all countries. www.who.int/ceh/indicators/en/

Percentage of children living in counties where air quality standards were exceeded

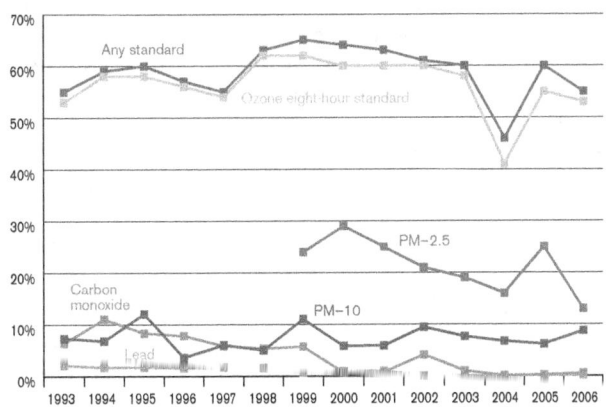

PM-10: Particles less than 10 micrometers in diameter

PM-2.5: Particles less than 2.5 micrometers in diameter

SOURCE: U.S. EPA, America's Children and the Environment www.epa.gov/envirohealth/children

DATA: U.S. EPA, Office of Air and Radiation, Aerometric Information Retrieval System

ENVIRONMENTAL HEALTH DISPARITIES

Eliminating health disparities, the gap in morbidity and mortality between social groups (e.g. racial/ethnic minorities and low-income populations), is of great importance to promoting child health. Current research suggests that characteristics of social, physical, and built environments contribute to these disparities. For example, children in socioeconomically disadvantaged communities are disproportionately diagnosed with asthma. In fact, black children in America are twice as likely to be hospitalized for asthma and four times as likely to die from asthma as white children. Eight percent of Hispanic children have asthma, but there are big differences among ethnic groups. Puerto Rican children have the highest rates of asthma at 20 percent, compared with 7 percent for Mexican-American children.

Southeast Pediatric Environmental Health Specialty Unit (PEHSU): The PEHSU conducted a research symposium on environmental health disparities in children. Break The Cycle is a collaborative, interdisciplinary research and training program with faculty from southeastern universities mentoring students to develop projects that reduce or prevent environment-related illnesses. The program addresses issues such as obesity, health effects of agricultural pesticides on children of migrant farmers, childhood asthma, disparities in girls' pubertal development, and environmental justice awareness. Students in diverse areas of study explore problems and solutions related to environmental health disparities. One case study on PCB contamination in Anniston, Alabama, included a biology seminar with 150 students at Spelman, a historically black women's college in Atlanta. The research was presented at a workshop in May at Emory University's Rollins School of Public Health. www.sph.emory.edu/PEHSU

Fact Sheets: Tools are needed to identify children who are most at risk and to understand the cumulative impact of social and physical environments on children's health. Feedback from users of EPA's *America's Children and the Environment* shows that information targeted for specific audiences is more effective than information aimed at general audiences. EPA's new fact sheets reflect this by addressing disparities in secondhand smoke exposure and asthma among African-American and Hispanic-American children. Each fact sheet includes information on actions that parents can take to protect their children, and positive actions EPA and other organizations are taking to address each environmental health issue. The intended audiences for these outreach materials are parents and community organizations working on environmental health issues for specific minority populations. These new fact sheets can be found at http://yosemite.epa.gov/ochp/ochpweb.nsf/content/publications2.htm#2.

LESSONS LEARNED AND THE CONTINUED NEED FOR RESEARCH

A Decade of Research: In March 2008, the National Center for Environmental Research (NCER) in EPA's Office of Research and Development released the

report *A Decade of Children's Environmental Health Research: Highlights from EPA's Science to Achieve Results (STAR) Program.* The report summarizes research from the program over the past 10 years, highlighting scientific findings in epidemiology, exposure science, genetics, community-based participatory research, interventions, statistics, and methods. This body of work has influenced policy and scientific research on children's health in the United States and abroad. The report is available at http://es.epa.gov/ncer/ publications/research_results_synthesis/.

Translating Research Into Practice: EPA, the National Institute for Environmental Health Sciences, the PEHSUs, and the Children's Environmental Health Centers organized a workshop for a group of clinicians, researchers, and health advocates from academia, government, and nonprofit organizations to discuss ongoing research in children's environmental health, issues in clinical practice, and opportunities for translating scientific findings. Topics included:

- Research on biomarkers for children's exposure to pesticides;

- Pesticide exposures and neurodevelopment in children from farmworker families;

- IPM in urban housing;

- Evidence concerning children's exposure to phthalates (chemicals in many plastics) as potential endocrine disruptors;

- Environmental management for asthma care and prevention;

- Outcomes of early life exposures to metals and neurotoxins;

- Transportation, the built environment, and children's health;

- The National Children's Study (NCS);

- Tools for recovery and communicating risks to children in the aftermath of disasters such as Hurricane Katrina; and

- The National Forum on Children and Nature, which is sponsoring demonstration projects addressing the issue of children's isolation from nature.

Conference Summary, Agenda, Presentations and Proceedings Report: http://es.epa.gov/ncer/childrenscenters/presentations/10_10_07/10_10_07_workshop.html

Environmental Factors and Puberty: Five articles based on the findings of the workshop entitled "The Role of Environmental Factors on the Onset and Progression of Puberty" were published as a supplement to *Pediatrics* on Feb. 1, 2008. This workshop, sponsored by EPA, NIEHS, and Serono International, was the first effort to evaluate the data and come to consensus across

disciplines, perspectives, and biases on the evidence for a secular trend in puberty timing (i.e., change in the age of puberty over time) and the role of environmental factors in puberty timing in children. The articles review the expert panel conclusions and the state of the science on the role of body fat, endocrine disruptors, and other environmental chemicals in puberty onset and progression from both the animal and epidemiological literature. Several EPA scientists are authors. http://pediatrics.aappublications.org/content/vol121/Supplement_3/

Food Allergies: Approximately 6–8 percent of children suffer from a food allergy during their first three years of life. Many of these children go on to develop a tolerance, and so the prevalence of food allergies in adults is approximately 3 percent. The reasons why some children outgrow the disease and others do not are unclear. As with other types of allergic diseases, the incidence of food allergy appears to be increasing and allergic individuals can experience life-threatening reactions. EPA has developed two complementary mouse models for food allergy that suggest thresholds exist for induction of these allergic responses. These models provide a means to explore underlying mechanisms for allergies and to understand the basis for susceptibility. http://toxsci.oxfordjournals.org/cgi/content/full/102/1/100

Children's Inhalation Dosimetry and Health Effects for Risk Assessment: Understanding children's risks due to exposure via the inhalation pathway is a special challenge. Children are different from adults and could experience different exposures due to differences in dosimetry (i.e., child-adult differences in respiratory/systemic dose resulting from exposure to the same airborne concentrations), child-specific behavior, and different health effects based on their growth and development. A special issue of the *Journal of Toxicology and Environmental Health* (vol. 71, no. 3, 2008) includes summary manuscripts from the children's

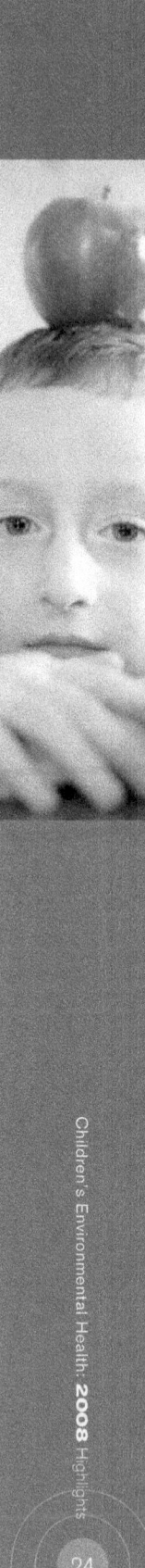

inhalation workshop that EPA hosted in 2006, as well as independent research articles on related topics contributed by workshop participants. This effort is the first to bring together the existing and emerging information on children's inhalation dosimetry, health effects, and risk assessment, and to explore and discuss new approaches for children's inhalation risk assessment practice.

Toxicity Testing: This past year, the National Academy of Sciences (NAS) released *Toxicity Testing in the 21st Century: A Vision and a Strategy*, funded in part by EPA. The report discusses how toxicity testing is poised to take advantage of advances in toxicogenomics, bioinformatics, systems biology, epigenetics, and computational toxicology to transform toxicity testing. The NAS report provides a blueprint for this paradigm shift.

National Children's Study (NCS):

Patterns of illness in children have changed substantially in the last 100 years. Infectious disease and sanitation used to be major risks to children's health. Today it is not uncommon to know a child with a chronic condition such as asthma, autism, or obesity. The extent to which environmental factors can influence the incidence of these diseases is not well understood. The NCS will answer important questions related to the environment and health, and will provide the foundation for government policy, medical practice, and individual decisions that will improve the overall health of children. The NCS will look at how the environment interacts with genetics and other factors to influence the health and development of 100,000 children in 105 locations across the United States. Children included in the study will proportionately represent the geographic, ethnic, and economic diversity of the nation's children.

The NCS is a result of collaboration by four lead federal agencies: the Eunice Kennedy Shriver National Institute of Child Health and Human Development (the study's home at the National Institutes of Health), EPA, CDC, and

NIEHS. This interagency coordination ensures alignment with the Children's Health Act of 2000, which authorized the study.

Much progress has been made in the past eight years. Twenty-two new Study Centers were announced to support the NCS in 26 communities in 2007; more will be announced in fall 2008. The first Study Centers are planning to enroll participants in 2009. Recent reviews of the NCS Research Plan by NAS and scientists in the lead agencies concurred that the study goals and design are responsive to the Children's Health Act. These reviews provide a number of recommendations, some of which are being implemented and others that can be considered in the future. www.nationalchildrensstudy.gov

International Childhood Cancer Cohort Consortium (I4C):

Childhood cancer is a devastating disease. Fortunately, the number of children with the disease is so small that a very large sample size is required to study it. Around the world, several large infant and child prospective studies have been launched to examine environmental and biological determinants of common diseases. The I4C was established in 2005 as a global alliance of longitudinal studies to enable investigations of the role of environmental exposures in the etiology of childhood cancer. Because of its longitudinal design and large sample size, it will be easier to see associations considered statistically meaningful. Initially, this effort could provide insights about the causes of childhood leukemia, and later could be helpful for studying other types of cancer as well as other rare childhood diseases. Protocols have been developed to examine the feasibility of combining data from studies from around the world. EPA supported two workshops, in 2005 and 2007. See more at www.nationalchildrensstudy.gov/.

PEDIATRIC ENVIRONMENTAL HEALTH SPECIALTY UNITS

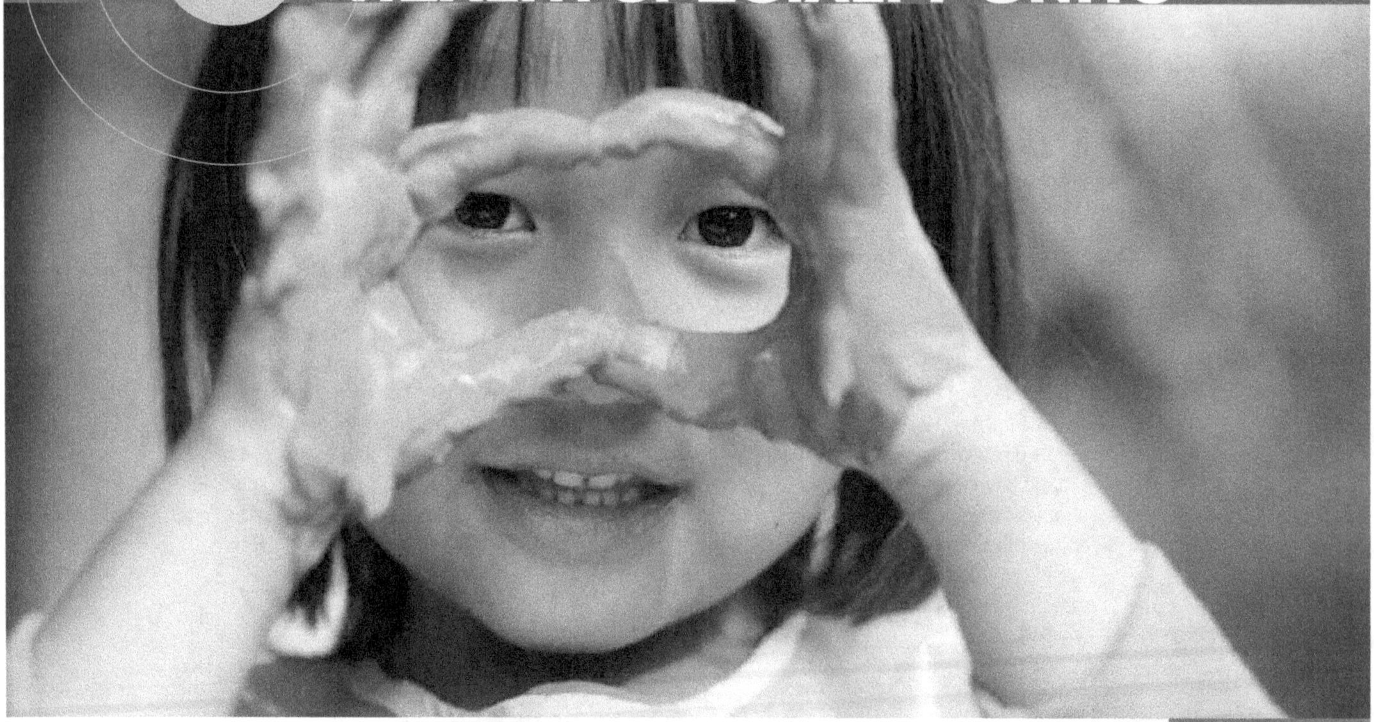

PEHSU
Pediatric Environmental Health Specialty Units

A unique partnership between EPA and the Agency for Toxic Substances and Disease Registry (ATSDR) has completed its 10th year. Both agencies support the PEHSU program, which provides education and consultation to health care providers, officials, and the community through their network of more than 100 physicians, nurses, toxicologists, and social workers in the United States, Canada, and Mexico. PEHSUs are a source of information and advice on environmental conditions that influence children's health. PEHSUs are academically based, typically at university medical centers, and work together as a network capable of responding to requests regarding prevention, diagnosis, management, and treatment of childhood conditions related to environmental factors. The need for pediatric environmental health expertise exists worldwide, and the PEHSU model is being adopted by other countries. Highlights from the PEHSUs include creation of fact sheets on wildfires and bisphenol A; participating in a Spanish webcast on lead poisoning; a successful effort to limit exposure to secondhand smoke in Tyler, Texas; working with colleagues in Chile to assist in creating a PEHSU in Santiago; addressing issues in child-care centers by working with the Children's Environmental Health Network; and creating a satellite PEHSU in Cincinnati, Ohio. www.pehsu.net

ENVIRONMENTAL HEALTH FOR HEALTH CARE PROVIDERS

Many children's health problems resulting from environmental exposures can be prevented, and need to be diagnosed, managed, and treated. While the public looks to health professionals for this, most are not educated on environmental precipitants of disease. The Institute of Medicine published two studies in the 1990s recommending that a greater effort be made to incorporate environmental health concepts into the training of health professionals.

Prenatal Environmental Health: EPA recently awarded grants to address environmental health issues during the prenatal period. Grantees will educate pregnant women about environmental health risks, demonstrate the effectiveness of information dissemination and behavior change, and increase the number of health professionals fluent in prenatal environmental health issues.

- The Inter-Tribal Council of Michigan, Inc. will address tobacco smoke, mercury, lead, and drinking water contaminants to Native American women of child-bearing age.

- The Oregon Department of Human Services aims to increase knowledge and promote behavior change among pregnant women who are exposed to mercury, lead, secondhand smoke, chemicals, pesticides, contaminated drinking water, and work- and hobby-related environmental hazards.

In Pennsylvania, the Philadelphia Department of Health will develop training and education programs for health care providers and pregnant women on environmental health risks of prenatal exposure to secondhand smoke and lead poisoning.

The Ohio Department of Health will develop and implement an easy-to-use environmental risk profile focusing on second-hand smoke, lead, mercury, radon, carbon monoxide, indoor pesticides, and other environmental toxics that can adversely affect birth outcomes.

The Duval County [Florida] Health Department will develop training programs for physicians, prenatal care providers, and Healthy Start staff on environmental health risks of prenatal exposure to methylmercury, lead, secondhand smoke, and drinking water contaminants.

Health Care Provider Grants: More than 9,000 health professionals were trained to address environmental health concerns as a result of projects funded by EPA. The University of Massachusetts Lowell educated health professionals with a combined client base of 60,000 children in the six New England states. Positive changes occurred in individual professional practice and organizational policy in these states, including nurse-led efforts to enact anti-idling legislation and incorporation of environmental health into public health nurse education. Another effort, by the National Center for Healthy Housing, brought together housing and health officials.

Other grantees include:

Canadian Institute of Child Health in partnership with the Asociación Argentina Médicos por el Medio Ambiente and Argentine Society of Doctors for the Environment

National Environmental Education Foundation

Greater Boston Physicians for Social Responsibility

International Pediatric Association

KSOHIA Collaboration of medical and public health universities in Kansas, Ohio, Iowa, and six Eastern European countries.

These grants support education and training of health professionals on environmental health issues. As a result of the train-the-trainer approach and the online tools developed, incorporating environmental health into health care provider education and practice will continue to expand beyond the life of these grants.

CHILDREN'S ENVIRONMENTAL HEALTH EXCELLENCE AWARDS

 Children's Environmental Health
2007 Champion

Dr. Ruth Etzel, an internationally known epidemiologist, pediatrician, and environmental health specialist, was the 2007 Children's Environmental Health Champion. Dr. Etzel was recognized for 20 years of work and for driving the effort to emphasize the critical importance of children's environmental health for health professionals. She founded and served as editor of the first and second editions of the American Academy of Pediatrics' *Pediatric Environmental Health.* Working with the Ambulatory Pediatric Association, Dr. Etzel helped launch the first Pediatric Environmental Health Fellowships in the United States in 2001. Most recently she worked with the International Pediatric Association to launch a virtual International Pediatric Environmental Health Leadership Institute.

2007 Excellence Award winners:

- The Asociación Argentina Médicos por el Medio Ambiente , on behalf of the International Society of Doctors for the Environment, Building Children's Environmental Health Capacity Among Health Care Professionals in the Southern Cone, Buenos Aires, Argentina

- Association of Occupational and Environmental Clinics, PEHSU program, Washington, DC

- Asthma Regional Council of New England, a program of the Medical Foundation of Boston, Environmental Investments Project, Dorchester, MA

- Children's Hospital of Pittsburgh Pediatric Residency Program, Pediatric Environmental Health Curriculum, Pittsburgh, PA

- Los Angeles Unified School District, Office of Environmental Health and Safety, Safe School Inspection Program, Los Angeles, CA

- Magee-Women's Hospital of the University of Pittsburgh Medical Center, Magee Environmental Health Initiatives, Pittsburgh, PA

- National Center for Healthy Housing, Pediatric Environmental Home Assessment Online Training, Columbia, MD

- Northwest Pediatric Environmental Health Specialty Unit, University of Washington, Seattle, WA

- Procter & Gamble Company, P&G Live, Learn and Thrive/Children's Safe Drinking Water, Cincinnati, OH

- EPA, National Exposure Research Laboratory, Research Triangle Park, NC

Children's Environmental Health: **2008** Highlights

29

CONCLUSION

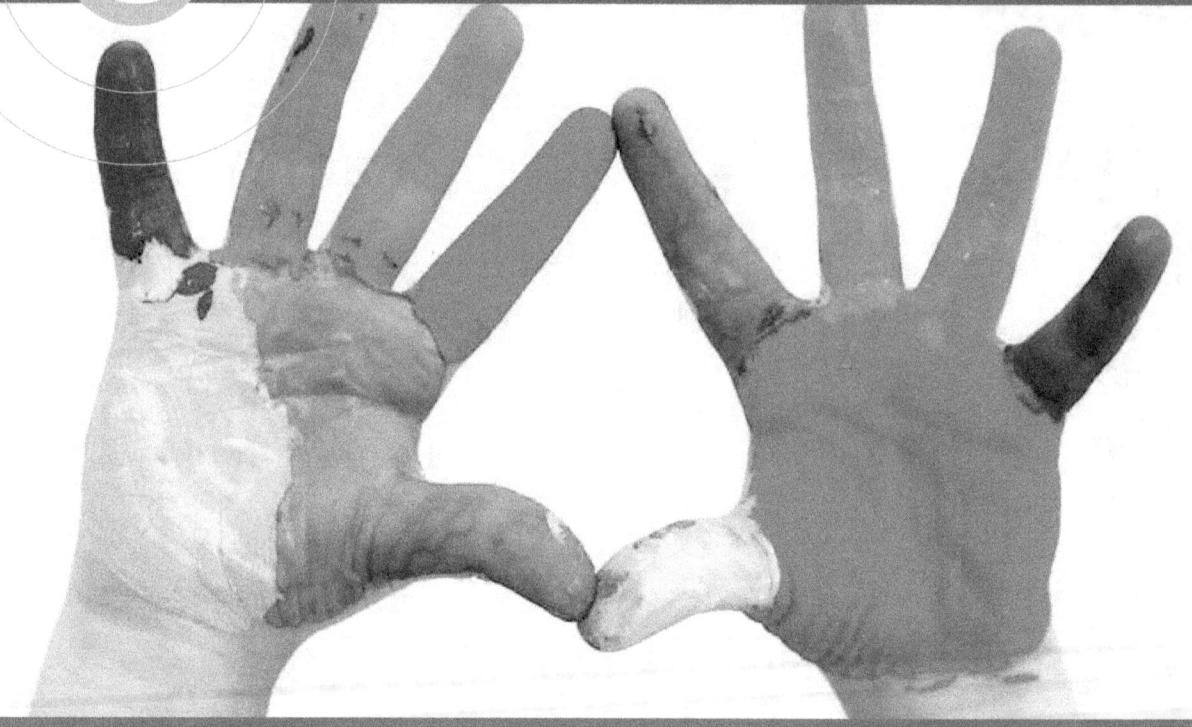

The environment clearly affects the health of children. Some environmental health issues are daunting in their scope, such as the effects of global climate change on the health of the world's children. Some problems mock us with their persistence, such as the disproportionate effects of exposure on minority and poor children. Some challenges are inevitable, such as natural disasters and their environmental health consequences. These concerns, together with the longstanding mission to clear our air, filter our water, restore the land, safely grow our food, remove waste, and treat sewage, require our continued commitment and collaboration with many diverse partners. EPA will continue to safeguard the health of children through its efforts to develop sound science, issue protective regulations, and raise awareness to create a healthy environment so that current and future generations continue to thrive.

Office of Children's Health Protection and Environmental Education
www.epa.gov/children

www.ingramcontent.com/pod-product-compliance
Lightning Source LLC
Chambersburg PA
CBHW081306170526
45165CB00011B/3434